PIGMANIA

PIGMANIA

EMIL VAN BEEST

FARMING PRESS
Wharfedale Road, Ipswich, Suffolk IP1 4LG

First published 1987
Reprinted, 1988, 1992.

British Library Cataloguing in Publication Data

Beest, Emil van
Pigmania.
1. Dutch wit and humor, Pictorial
I. Title
741.5'9492 NC1549

ISBN 0-85236-167-X

Cover design: Hannah Berridge

Printed and bound in Great Britain by
Page Bros, Norwich

1
2
3
4
5
6
1
2
3
4
5
6

RHINITIS
RESEARCH
VET

1
JAN
31
DEC

REMEDY FOR
TAIL-BITING

AUSTRALIA
ENGLAND
HOLLAND

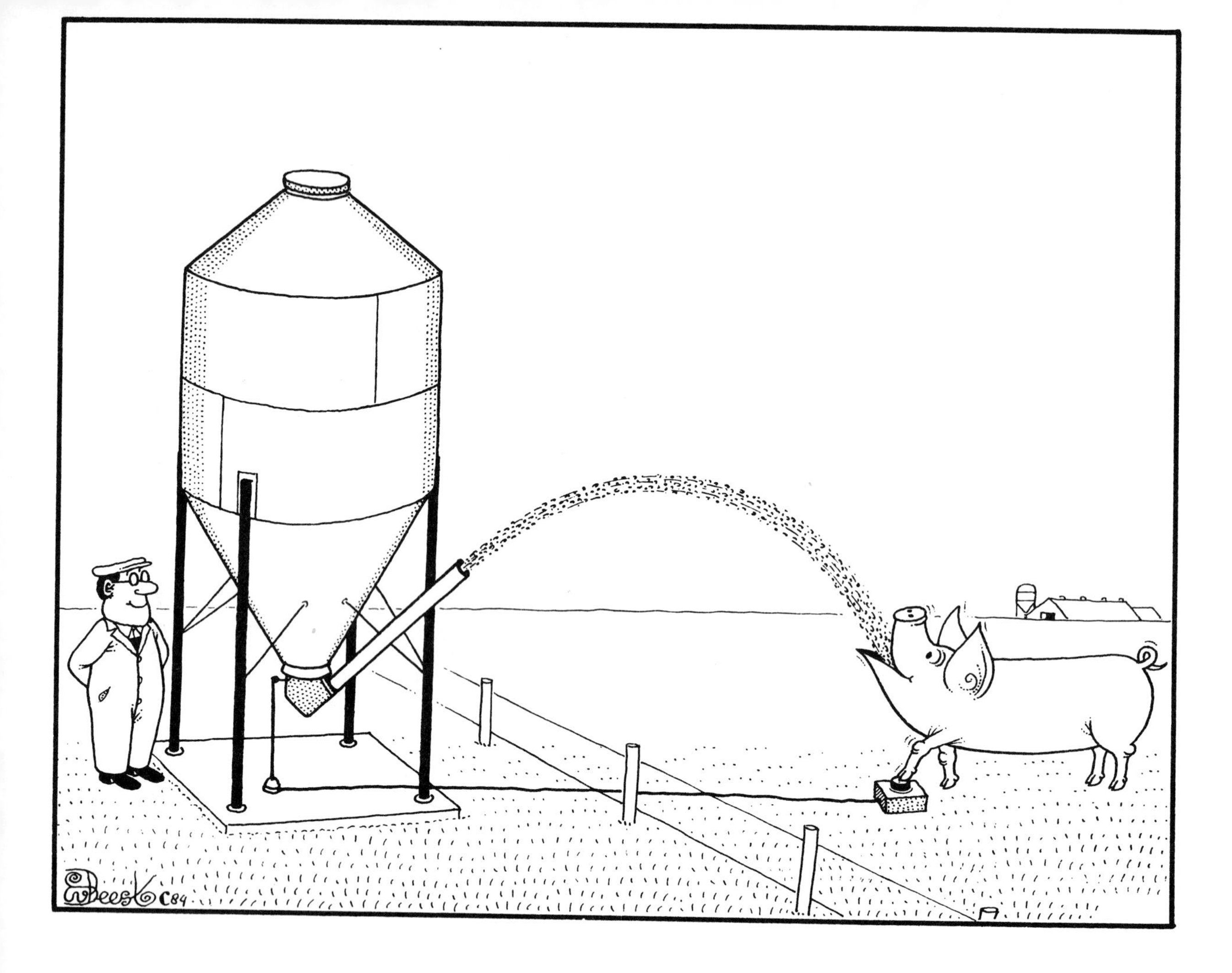

RUBBER TAILS
STOPS TAIL-BITING

EN5387

How to
prevent
tail biting

MATINÉE

Emil van Beest comes from Goor, in the Netherlands. His cartoons have been a regular feature of the Dutch agricultural magazine, *Boerderij*, since 1965 and have appeared throughout the world.

This collection is drawn mostly from those which have been published in Britain in *Pig Farming* magazine.

FARMING PRESS BOOKS AND VIDEOS

Farming Press publishes a wide range of books and videos about farming.

Humorous writers include Veronica Frater, Henry Brewis, John Terry and James Robertson. Farming in the past is catered for by Michael Williams' tractor books, as well as Michael Twist's ***Spacious Days***.

The range of practical books and videos for farmers and students published by Farming Press is unrivalled in Britain. The list includes titles on pig, sheep, dairy and arable farming as well as many other farming and veterinary topics.

For a free illustrated catalogue please contact:

Farming Press, Wharfedale Road, Ipswich IP1 4LG.